Today's Space Elevator Assured Survivability Approach for Space Debris

Peter Swan, Ph.D.
Michael Fitzgerald
Cathy Swan, Ph.D.

Prepared for the
International Space Elevator Consortium
Chief Architect's Office

March 2020

International Space Elevator Consortium   ISEC Position Paper # 2020-1

International Space Elevator Consortium   *ISEC Position Paper # 2020-1*

Today's Space Elevator Assured Survivability Approach for Space Debris

Copyright © 2020 by:

Peter Swan
Michael Fitzgerald
International Space Elevator Consortium

All rights reserved, including the rights to reproduce
this manuscript or portions thereof in any form.

Published by Lulu.com

info@isec.org

978-1-67818-191-8

Cover Illustrations:
Front – Image from NASA Orbital Debris Office
    https://orbitaldebris.jsc.nasa.gov/images/beehives/leo1280.jpg

Printed in the United States of America

International Space Elevator Consortium　　　*ISEC Position Paper # 2020-1*

Preface

The Space Elevator is a Catalyst for Change!

Space debris is expected to be a part of space operations for most of this century. The real mitigation approach is the establishment of policy and actions that will prevent, and extensively reduce, creation of debris in the first place. The Space Elevator must become a catalyst to instigate more aggressive and active removal and mitigation of space debris. This position paper is the International Space Elevator Consortium's (ISEC) attempt to document an approach to mitigate the Space Elevator mission impact and identify safety issues when space debris threatens Space Elevator tethers. We implore other organizations and activities to also confront this issue.

ISEC believes that debris mitigation concepts will be built, operating, and thriving well before a Space Elevator Transportation System reaches operational status. To that end, this paper serves as the initial characterization of this transportation system which can identify the needed performance for debris mitigation systems.

The current topic of discussion within ISEC about space debris is how to work with others and develop programmatic and engineering solutions. Currently, the ISEC leadership is working within the following topics[1]:

>   ***Debris alert*** ➔ ***warning needs***
>   ***Debris sizing*** ➔ ***as a threat variant***
>   ***Space Elevator Tether Movement*** ➔ ***passive defense***
>   ***The Sentry System*** ➔ ***an architecture adjunct for active defense***
>   ***System Recovery*** ➔ ***post debris-event actions***
>   ***Improving the Baseline*** ➔ ***configurations to enhance mitigation***

The entire matter will remain active in the space elevator system's risk management program. Multiple approaches will be implemented to include:

---

[1] Fitzgerald, M., "Space Elevator Architecture's Architecture Note #25, Debris Mitigation Roles, ISEC note, www.isec.org, March 2019.

- Direct, real time coordination will be established with the military's Combined Space Operations Control Center [CSpOC]
- Creation of a "Debris Assessment Chair" at the Space Elevator Operations Center with unrestrained involvement in daily operations planning and execution.
- As might be necessary, improved Debris Mitigation activities will be a mandatory inclusion in all system calls for improvement --- (See Architecture Note #29 "Call for Improvement" Policy.)

This position paper will address these topics inside a summary of concerns and discussions on future global approaches towards the mitigation of space debris.

International Space Elevator Consortium *ISEC Position Paper # 2020-1*

Acknowledgements

This document could not have been accomplished without help from the entire International Space Elevator Consortium (ISEC) team, who have been dreaming "big" for years now. I would like to recognize the traits of our Board (and other passionate space elevator supporters) as described by Gino Wickman as "Six essential characteristics of entrepreneurial types."[2]

1. Visionary: Perhaps one of the most innate trait, and therefore, the most difficult to teach is vision.

2. Passionate: A truly successful entrepreneur must be passionate about one's products and services to the point that it fills an almost existential void. This involves an unshakeable belief in the value and necessity of one's venture— one that fulfills and grants purpose.

3. Problem-solver: Entrepreneurship is innate: it is a trait of nature, not nurture. Challenges are exciting opportunities, not setbacks, and an entrepreneur's natural response is optimism rather than discouragement.

4. Driven: The best entrepreneurs fuel themselves with an intrinsic motivation to competitively overcome failure after failure - matching enthusiasm with endurance and a sense of urgency.

5. Risk-Taker: In many ways, risk aversion is the enemy of entrepreneurship. Another hardwired trait, a risk-loving mentality, is key to rapid iteration, failing forward, and an entrepreneur's investment in seemingly crazy ideas that revolutionize entire industries.

6. Responsible: At the same time, however, entrepreneurs must be responsible in both risks taken and venture management. Always remaining accountable.

As I read through this list, I recognized that our board of directors has many of these traits with some running across them all. I think two that are dominate within the ISEC are Visionary and Passionate.

Thanks are due to those who have contributed. John Knapman, Michael Fitzgerald, and Dennis Wright for their inputs and review of the document. I also want to especially thank Dr. Cathy Swan who reads every word (several times) as an editor translating engineering into English. Well done Space Elevator team!   Peter Swan, Ph.D.

---

[2] I am quoting Peter Diamandis' email blog, [Diamandis, Peter, "Entrepreneurs-in-the-Making: Are you Ready to Take the Leap?", email 19 Dec 2019] from a book "Traction" by Gino Wickman.

International Space Elevator Consortium　　　*ISEC Position Paper # 2020-1*

Executive Summary

Space Debris is Manageable for Space Elevators

The International Space Elevator Consortium's (ISEC) position has been well documented and discussed. The space elevator activities about space debris were initiated in the 2010 ISEC Study Report, "Space Elevator Survivability, Space Debris Mitigation" after a full year of analyses by space debris and space systems experts. Since then, there have been events that have increased the growth of space debris. This 2020 report has taken a look at the situation and extrapolated across the arena to arrive at some preliminary results. The numbers were calculated for the present (2019 tracked debris data), compared to the past (2010 data), and the future (2030 estimates with projections of new satellite constellations). The approach, as discussed in the 2010 space debris report, is one where the volume of space around the Earth is shown to have a density of debris related to altitude zones. That report breaks out the zones, analyzes the information and drives conclusions. The collision probability analyses are linear with respect to numbers of debris within the volume occupied by a 100,000 km of one-meter wide tether. The real efforts focused upon high debris density regions with identified zones between 200 and 2000 km altitudes. The report takes the density numbers, extrapolates the probabilities of collision and arrives at conclusions. The Executive Summary of 2010 Report stated: "To assess the risk to a space elevator, we have used methodology from the 2001 International Academy of Astronautics (IAA) Position Paper on Orbital Debris[3]:

> The probability (PC) that two items will collide in orbit is a function of the spatial density (SPD) of orbiting objects in a region, the average relative velocity (VR) between the objects in that region, the collision cross section (XC) of the scenario being considered, and the time (T) the object at risk is in the given region."

---

[3] 2001 Position Paper On Orbital Debris, International Academy of Astronautics, 24.11.2000. download for free from www.isec.org

ix

$$PC = 1 - e^{(-VR \times SPD \times XC \times T)}$$

Using this formula, we calculate the Probability of Collision for Low Earth Orbit (LEO), Medium Earth Orbit (MEO), and Geosynchronous Orbit (GEO). Our focus is on LEO -- as fully two thirds of the threatening objects are in the 200-2000 km (LEO) regime. Our analyses show: Space Debris can be reduced to manageable levels with relatively modest design and operational "fixes."

With discussions and calculations across three decades, the conclusion stays the same: for time periods - 2010, 2019 and 2030.

"Space debris mitigation is an engineering problem with definable quantities such as density of debris and lengths/widths of targets. With proper knowledge and good operational procedures, ... space debris is not a show-stopper by any means. However, mitigation approaches must be accepted and implemented robustly."

With the realization that there is much to do in architectural and engineering approaches to space debris mitigation, the following concepts have been assessed as first approximations:

1. Architectural and Engineering Design Inputs:
   - Multi-leg design
   - Designing the tether itself to survive small debris hits
   - Include a repair tether climber that mends small holes or rips in tethers
   - Support operational approaches shown below.

2. Operational Approaches:
   - Passive Approaches for Debris Mitigation: multi-leg design, varying tether design by altitude, and multiple parallel tethers for greater carrying capacity
   - Active Approach: tether movement upon demand, on-orbit Sentry Satellite System, and approach for recovery from tether sever

3. Collaboration with Others:
    - Establish operational co-operations with Space Traffic Management organizations
    - Coordinate with owners of space assets (especially derelicts) at GEO
    - Coordinate with organizations who will remove space debris
    - Establish operational procedures to receive timely warnings and then respond to them

4. Timely Debris Alert & Warning:
    - ISEC foresees a close and interactive communication with the military Combined Space Operations Control Center - known familiarly as CSpOC. CSpOC is responsible for tracking thousands of debris pieces and providing orbital parameters of those pieces to space operational users. In addition, commercial capabilities have emerged which offer forming and formatting of that information - operationally satisfying their commercial customers.
    - Projecting future collisions is an important portion of the tasking for CSpOC, enabling timely warning of predicted conjunctions to be sent to the space elevator operations center. This timely warning should enable actions to move portions of the tether to avoid those predicted conjunctions.
    - ISEC expects that the space elevator system operator will be able to depends upon a warning forecast within 72 hours of a convergence / close approach to a Space Elevator tether location. The Space Elevator team expects that CSpOC will hold a position as the Debris Mitigation chair in the Space Elevator Operations Center.

Space Debris can be managed!

Table of Contents

Preface v
Acknowledgements vii
Executive Summary ix
Table of Contents xiii

Chapter 1: Introduction 1

Chapter 2: Architectural Challenge 4

Chapter 3: Survivability Design Inputs 12

Chapter 4: Collaboration 15

Chapter 5: Operational Approaches 17

Chapter 6: Conclusions for 2030 22

Chapter 7: Recommendations: 24

Appendices: 27
Appendix A: 2010 Study Conclusions 28
Appendix B: 2010 Study Recommendations 30
Appendix C: Space Elevator Architecture Note #25, March 2019 33
Appendix D: ISEC Completed Studies 37
Appendix E: Description of International Space Elevator Consortium (www.isec.org) 43

Chapter 1: Introduction

The historical growth of space debris is an issue across the full space mosaic. The rapid increase of items in space demands that there be efforts across the industry to reduce these numbers as space missions are expanding to constellations of satellites. The good news is that there are several companies and countries addressing active reduction as we enter the third decade of the 21st Century (actually entering the eighth decade of spaceflight). This position paper will address the whole arena of space debris and future space elevators. Early inputs to the architectural approach for engineering and operational challenges are important and should be evaluated early in the design processes. Space debris is not a show-stopper; but, it must be incorporated into the engineering design. An ISEC initial year-long study was conducted from 2009 to 2010 addressing this exact issue. The title of the study was "Space Elevator Survivability Space Debris Mitigation."[4] [download pdf free version at www.isec.org) This 2020 position paper shows that the increase in space debris numbers across 21 years has not altered the results of the study (numbers of space debris for 2009, 2019 & 2030 are NASA supplied). In the preface of the first study report, the President of ISEC (Ted Semon) was quoted as saying:

> "This study represents the culmination of efforts by the contributors and answers the question: Will space debris be a "show stopper" for the development of the Space Elevator Infrastructure?
>
> The answer is a resounding NO!

The recognition of space debris risk with reasonable probabilities of impact is an engineering factor. The proposed mitigation concepts change from a perceived problem to a concern; but, by no means is it significant. This study illustrates how the development office for a

---

[4] Swan, P., R. Penny, C. Swan, "Space Elevator Survivability Space Debris Mitigation," Publisher Lulu.com, 2010.

future space elevator infrastructure can attack this problem and convert it into another manageable engineering factor."[5]

This current ISEC Position Paper will address major topics focused upon space elevator operations, management and survivability. The topics to be expanded upon are:

- Architectural Challenge: Design challenges including analyses and understanding debris density and distribution.

- Collaborations with Space Traffic Management (STM) activities and organizations: Space Elevators need to provide inputs to STM design efforts, monitor progress, conduct close co-operational activities, and ensure avenue for active debris warnings.

- Operational approaches to space debris:
    - Passive - multi-leg operations, different tether designs vs. altitude
    - Active - tether movement on-demand, on-orbit Sentry System for aggressive operational protection and recovery from a severance - a low probability event.

---

[5] Swan, P., R. Penny, C. Swan, "Space Elevator Survivability Space Debris Mitigation," Publisher Lulu.com, 2010.

Chapter 2: Architectural Challenge

2.0 Background: The expertise of the Chief Architect and Chief Engineer for Space Elevators will feed into discussions on "how to" execute the program. The beauty of the developmental process is that very knowledgeable professionals gather together and analyze thousands of available options at the beginning of a mega-project. The future will be defined with this initial set of evaluations. In the early analyses on "how to" build a space elevator, concepts develop dealing with how to manage threats from space debris. Debris was discussed ten years ago as ISEC's first study as the team felt the need was evident and should be analyzed and presented to those in leadership positions. This update is also in response to this understanding of the need to explain.

2.1 Design Challenge - Threat Analysis:

Table 2.1 Satellite Box Score (2010)

Satellite Box Score			
(as of April 7 2010, as catalogued by the U.S. Space Surveillance Network			
Country/ Organization	Payloads	Rocket Bodies & Debris	Total
China	85	3207	3292
CIS	1400	4370	5770
ESA	38	44	82
France	48	421	469
India	39	131	170
Japan	112	77	189
US	1127	3694	4821
Other	463	114	577
Total	3312	12058	15370

2.1.1 Space Debris Growth: The space elevator discussion about space debris was addressed in the 2010 ISEC Study report after a year of analyses by space debris and space systems experts using the NASA information in table 2.1 Satellite Box Score (Apr 2010). ISEC has extrapolated across the space debris arena. The present analyses started with numbers from 2010, recognized the growth to 2019, and then predicted growth over the next ten years. This report breaks out the altitude zones, analyzes the altitude density numbers and summarizes its conclusions. These analyses are linear with respect to probability of conjunction of debris with a 100,000 km tether, one-meter wide (double the debris density, double the probability of collision). The real concern is for high-density regions with identified zones between 200 and 2000 km altitude. The probability of

SATELLITE BOX SCORE
(as of 04 October 2019, cataloged by the U.S. SPACE SURVEILLANCE NETWORK)

Country/Organization	Spacecraft*	Rocket Bodies & Debris	Total
CHINA	369	3720	4089
CIS	1536	5099	6635
ESA	90	57	147
FRANCE	66	507	573
INDIA	96	163	259
JAPAN	180	115	295
USA	1878	4815	6693
OTHER	966	122	1088
TOTAL	5181	14598	19779

* active and defunct

collision will then be estimated and extrapolated in order to reach conclusions.

Table 2.2, Satellite Box Score Oct 2019.[6]

The next step was to determine the 2019 numbers. As shown in the second table, the total number of space debris is 19,779 pieces, as of Oct. 2019. Comparison with the 2010 Space Elevator Survivability Space Debris Mitigation report shows a total Satellite Box Score debris larger than 10 cm in April 2010 was 15,370 while in October 2019 the number is 19,137. This 25% increase over 8 years is actual data shown in the NASA Orbital Debris Quarterly Reports.

A projection was then made towards 2030 space debris numbers. There were two approaches to this projection:
1. Simply double the NASA numbers, as a first approximation, as there will be many new constellations launched within the next ten years. This estimate is probably high as the number of constellations will be mitigated with the realities of launch, operations, and funding opportunities awaiting each proposed business model. (estimate: 38,274)
2. Or, use the NASA estimate, once again from the Quarterly magazine on space debris (38,000).

---
[6] Sep 2018 Orbital Debris Quarterly News (NASA Johnson Center office).

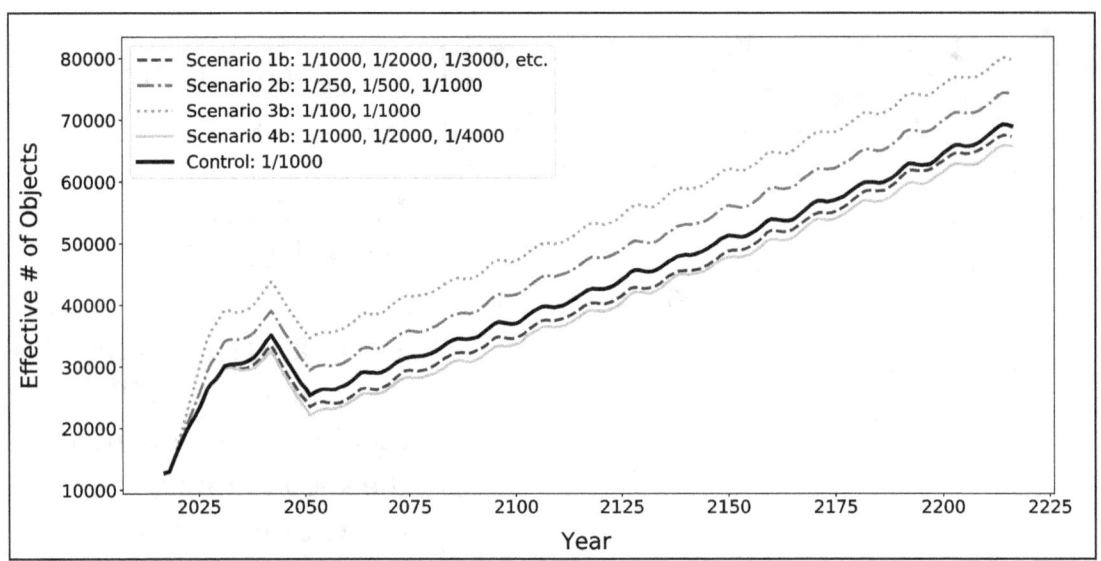

Figure 2. Effective number of objects projected to be in orbit after 200 years with varying explosion rates over each replenishment cycle. The bulge represents the constellations deploying and the subsequent fall-off represents the end of the constellations lifetime, i.e., there are no additional constellations being added to the environment.

Table 2.3, Projection of objects incorporating potential constellations.[7]

An estimate, leveraging the combined approach, results in a total number of debris objects in space in 2030 of 38,000, as shown in Table 2.3. This results in a spread of debris population, as shown in Table 2.4.

Item *(> 10 cm)*	*2010*	*2019*	*2030* *Est.*	*Estimated* *with*
Debris by NASA	15,378	19,137	38,000 38,274	NASA numbers 2019 doubled

Table 2.4, Time population of Space Debris

2.1.2 Space Debris Probabilities: The 2010 methodology was an approach based upon probabilities. In addition, to accomplish this, the breakout in altitude distribution had to be understood. Two of the conclusions from the 2010 study report showed that the densities at GEO and MEO were very small and manageable with monitoring and simple tracking (estimates to be slightly enhanced with numbers covering 2030). The densities at Low Earth Orbit were the ones that needed to be analyzed. One puzzle was the distribution of space debris with respect to altitude. As such,

---

[7] Orbital Debris Quarterly, NASA Orbital Debris Program Office, Aug 2019.

the NASA chart showing the numbers of debris vs. altitude was used as a baseline. It was estimated that the numbers have increased in LEO. The spread can still be used as a baseline distribution of densities for future calculations. The chart used in 2010 shows the following distribution.

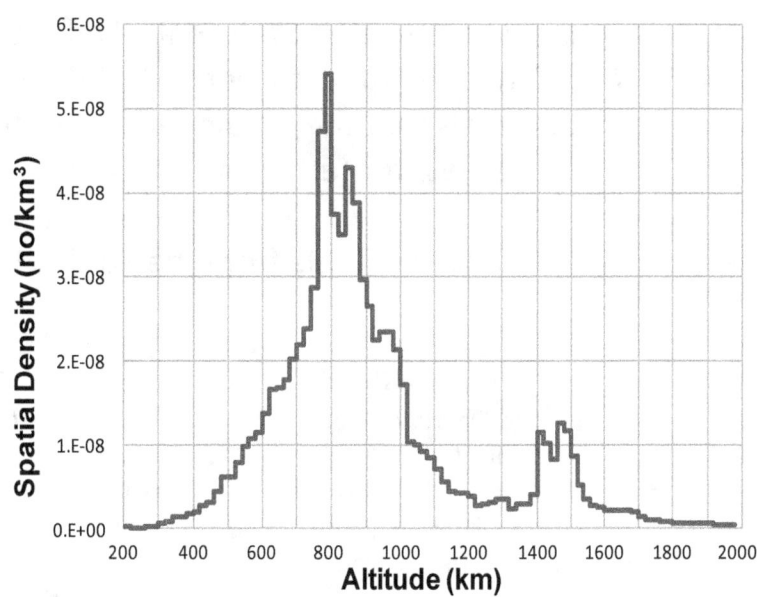

Figure 2.1 LEO Spatial Density (2010)[8]

Now that the distribution of space debris is shown, specific cases can be analyzed. As such, in LEO, there were three cases studied, with three variations. They were:

For tracked debris:
    Case A: 60 km ribbon segment (740-800 km altitude) representing the peak debris density – highest risk case.
    Case B: 60 km ribbon segment (1340-1400 km altitude) representing an average debris density in LEO.
    Case C: 1800 km ribbon segment (200-2000 km altitude) representing the entire LEO environment.

For un-tracked debris: (estimated as ten times tracked debris[9])
    Case A-u, B-u, C-u: represent the untracked items in above described segments.

For satellites under control (estimates at six percent of tracked debris[10])

---

[8] With permission from Debra Shoots, NASA Orbital Debris Program Office, May 2010.
[9] Estimate derived from discussions with space debris experts
[10] Estimates from discussions with space debris experts

Case A-c, B-c, C-c: represent the controlled satellites in above segments. Estimated to be six percent of the tracked debris. Significance here is that they could maneuver to avoid other debris or space elevator tether.

This is shown in the following table from the Study.

Types of Debris	Case	Comment
Untracked Debris < 10 cm		10 x tracked
60 km stretch - peak	A-u	Highest Density
60 km stretch - average	B-u	Average LEO
LEO 200 - 2000 km	C-u	Total LEO region
Tracked Debris > 10 cm		
60 km stretch - peak	A	Highest Density
60 km stretch - average	B	Average LEO
LEO 200 - 2000 km	C	Total LEO region
Cooperative Debris		0.06 x tracked
60 km stretch - peak	A-c	Highest Density
60 km stretch - average	B-c	Average LEO
LEO 200 - 2000 km	C-c	Total LEO region

Table 2.5 LEO Regional Breakout by Cases

Once the numbers have been allocated by altitude regions (resulting in volumetric densities), the next step could be taken - calculations of probable intersections. The Executive Summary of 2010 ISEC Report stated: "To assess the risk to a space elevator, we have used methodology from the 2001 International Academy of Astronautics (IAA) Position Paper on Orbital Debris."[11] The approach was described as:

"The probability (PC) that two items will collide in orbit is a function of the spatial density (SPD) of orbiting objects in a region, the average relative velocity (VR) between the objects in that region, the collision cross section (XC) of the scenario being considered, and the time (T) the object at risk is in the given region."

$$PC = 1 - e^{(-VR \times SPD \times XC \times T)}$$

---

[11] 2001 Position Paper On Orbital Debris, International Academy of Astronautics, 24.11.2000.

Using this formula, the Probability of Collision can be calculated for LEO, MEO, and GEO. Our focus is on LEO -- as over two thirds of the threatening objects are in the 200-2000 km (LEO) regime. Our analyses, as shown in the 2010 ISEC Study Report, concluded:

> ***The threat from Space Debris can be reduced to manageable levels with relatively modest design and operational "fixes."***

So now we layout the conclusions from the 2010 report and extrapolate to 2019 and 2030. This is done in a linear manner as this is a straightforward projection.

Item	*2010*	*2019*	*2030 Est.*	*Comment*
Total Tracked Debris by NASA (2010 & 2019 measured, 2030 estimated)	15378	19137	38,000	Assume Internet constellations will add many space objects by 2030
Threats in GEO region (possible conjunction)	0.0026 per year	0.005 per year	0.01 per year	Good operational procedures a must.
Threat in MEO region (possible conjunction)	0.0003 per year	0.0006 per year	0.0012 per year	Good operational procedures a must.
Untracked, small (<10 cm) debris will impact a Space Elevator in (LEO 200-2000 km), on the average;	Once every ten days	Once every 7.5 days	Once every 4 days	Design for tether and movement planned to account for this - with continuous repair[12]
Tracked debris will impact the total LEO segment (200 – 2000 km) if no actions are taken.	Once every 100 days or multiple times a year	Once every 75 days or several times a year	Once every 40 days or every two months or so	Note, this assumes there is no active movement of tracked objects or movement of the tether
Tracked debris will only impact a single 60 km stretch of LEO space elevator, on the average	Every 18 years with every 5 years in peak regions	Every 14 years with every 4 years in peak regions	Every 7 years with every 3 years in peak regions	Note, this assumes there is no active movement of tracked objects or movement of the tether

Table 2.6 Summary by Year and Altitude Region

---

[12] Repair of tether from small debris impacts is a must in the design, development and operational phases

As a result, the conclusion stays the same - for 2010, 2019 and 2030.

"Space debris mitigation is an engineering problem with definable quantities such as density of debris and lengths/widths of targets. With proper knowledge and good operational procedures, the threat of space debris is not a show-stopper by any means. However, mitigation approaches must be accepted and implemented robustly."[13]

---

[13] Swan, P., R. Penny, C. Swan, "Space Elevator Survivability Space Debris Mitigation," Publisher Lulu.com, 2010.

Chapter 3: Survivability Design Inputs

The roles of the future Chief Architect and Chief Engineer of the space elevator development team are to understand the developmental process and enable valuable inputs to ensure the design team has the needs and requirements outlined and understood. In the arena of space debris reduction, there are several design inputs that need to be instigated. They are broken into architectural level and engineering inputs for design considerations.

Architectural and Engineering Design Inputs: In this area of activity three concepts are upfront and necessary to be considered in the design process. They are Multi-Leg Architecture, tether design for small debris impact and potential penetration, and tether repair.

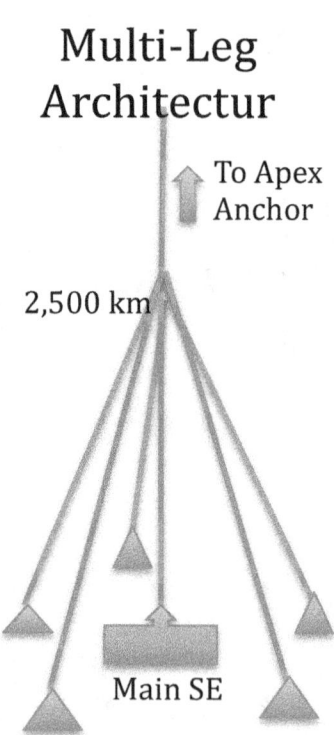

- The multi-leg design has the concept of having a principal leg for day-to-day operations of tether lift-offs and climbing. The other tethers are secondary in that they are there for backup in case of potential severance; but, they could be used to accomplish other missions such as low altitude hotel or scientific instrument placement. One concept is six legs with joining of the legs at the 2,500 kms altitude - above dense LEO debris belts.

Figure 3.1, Space Elevator Multi-Leg Design

- The second approach is designing the tether to survive small debris (< 10 cm in diameter). This has been discussed in many papers with the leading contender developed by Tethers Unlimited called the Hoyt Tether - a woven design spreading the tensile strength across multiple strands to ensure that if one is cut, the others share the load. Other tether designs, such as the use of multiple layers of a 2D material such as single crystal graphene, need to be examined and tested for the effects of ballistic penetrations (of small objects with great energy).

- The last topic is that of having a repair tether climber going up and down repairing small holes or rips in tethers from wear and tear or small debris penetrations. The current concept would put sensors on the front of each tether climber, inspecting as they go. Then a repair tether climber would be sent to the area of concern and patch or weave a "fix" for the tether for that location.

Chapter 4: Collaboration

Engineering Management Inputs: This area of activity will focus on the ability of the Chief Architect to coordinate with and request co-operation in the area of space debris on daily operations. Some of those cooperation agreements would be called adjunct to the design of the space elevator, but they are just as important as any element of the developmental process. Some of these would include:
- Establish cooperation with organizations operating the Space Traffic Management processes and day-to-day operations (with cross-personnel swaps to enhance operational knowledge).
    - This would include providing knowledge of space elevator locations (to include hourly locations of all 100,000 kilometer elements of the tether and projections for the next 24, 48, and 72 hours) and receiving warnings and "heads-ups" of potential conjunctions from tracked debris or operational systems.
    - Provide inputs to Space Traffic Management design teams to ensure space elevator requirements are satisfied.
- Coordination with owners of space assets (to include derelicts) that can be used by space elevator operations for Apex Anchor mass or construction of needed facilities at GEO.
- Work closely with those organizations responsible to remove debris from orbit. The elimination of major debris in orbit is critical to all travel and operations in space (this is a must - and there are many people around the world who believe debris removal should be started soon to mitigate future challenges to normal spaceflight - this is NOT simply a space elevator issue; but, it is important for all spaceflight).
- And, set up a rapid capability for the Space Traffic Management System to instantly notify the space elevator community if there is a projected conjunction.

These cooperative activities must be started early and be comprehensive in reach. Tracking of space debris and warning of potential conjunction are responsibilities of others, but so essential to space elevator success that early involvement is necessary to input special needs of space elevators.

Chapter 5: Operational Approaches

5.0   Various Approaches: The strengths of early involvement of space elevator architectural and engineering teams with Space Traffic Management operations is that preliminary approaches can be discussed, refined, built into both systems (SE & STM) and then operationally executed. Some of these ideas (split into passive and active approaches) are:

5.1   Passive Approaches for Debris Mitigation: A tremendous strength of the space elevator is that it is a permanent infrastructure and can be described and defined with precision. This enables the infrastructure to leverage multiple ideas, two of which are described below:

5.1.1 Multiple Leg Architecture: This was mentioned earlier as having as many as six legs from 2,500 km altitude to the surface of the ocean. As such, there would be some weight added to the mainline tether to be supported. Calculations show that it is well within the variable design of the tether mass. The strength of the concept is that if one leg is severed below this node, the other tethers share the load and the whole system can be reconfigured within a short time period. It would not significantly affect any tether climber operations below 2.500 km nor the operations at GEO or beyond. Of course, the dynamics of a multi-leg architecture will need to be studied. The recovery of operations would be somewhat routine.

5.1.2 Varying Tether Shape: Many requirements are out in the design process for the shape of the tether. Some of the ideas are:
- Round tether in the wind region to lower the cross section
- Curved ribbon one-meter wide with a woven design in the high debris threat region to ensure all the non-tracked small space debris "blows through." This tether design enables tensile stress to be carried by surviving threads. The numbers are such (as shown in previous sections of this position paper) that the odds of two such holes in any 60 kilometer stretch are every three (2030) to five years (2019).
- Multiple tethers sharing the load of each tether climber for redundancy. This is the assumed design when the Full Operational

Capability space elevator is opened in 2050 to ensure redundancy for human transportation and larger payloads (85 metric tons).
- In addition, the concept of multiple tether designs based upon altitude reflects the advantages of distance in that the gravitational force from the Earth decreases at an exponential rate ($1/r^2$). This leads to lower power requirements as the climb gains altitude to climb against gravity, thus changing design of tether.

5.2 Active Approaches for Debris Mitigation: The dynamics of a space elevator enable motion at any time and at any position along the tether. As such, many concepts have been developed that use this strength as well as enhanced motion when desired.

5.2.1 Tether Movement upon Demand: One of the ideas from the very earliest work on space elevators explains how tether dynamics lend themselves to motion off the direct line from Earth Port to Apex Anchor. By monitoring every element along the tether (assume a single element is 1 km), the prediction capabilities of fast computers should enable us to recognize an approach from space debris with a potential conjunction and then move that particular element of the tether out of the path. This requires two factors: (1) fast computing time monitoring each element to project future motion of that element and (2) precise threat warning on potential conjunctions of space elevator element and pieces of large space debris.

The instigation of "off-routine" motion of the tether can be accomplished with many techniques, such as:
- Reel-in/out from GEO, Earth Port, and/or Apex Anchor
- Slow down/speed up/stop and reverse direction of any of multiple tether climbers along the tether
- Movement at Earth Port, GEO and/or Apex Anchor

5.2.2 On-Orbit Sentry Satellite System: This concept is based upon the design of an electromagnetic tether satellite that can maneuver up, down, and across magnetic fields lines to rendezvous with any approaching space debris. In a concept called Sentry, space debris will be intercepted and removed. Debris headed for any portion of the Space Elevator will have intercept priority; but, if otherwise not encumbered by that priority the Sentry will gather and dispose of other space debris as a matter of course.[14] This Sentry System would then approach, attach itself, and move/deorbit the body. Multiple Sentries would be needed for each Space Elevator in different altitude ranges to ensure rapid responses.

5.2.3 Recover from Severance: This is the lowest probability event to plan for; but, it must be addressed systematically. The first assumption is the sever will occur at less that 2,000 km altitude - with the maximum likelihood at 800 km or 1,400 km altitude regions. These have the highest densities of space debris. The geosynchronous belt has less debris, a larger volume of operations and slower moving debris (derelict satellites). In addition, the mid-orbit region has significantly less satellite debris and a huge volume, with really fast debris. As such, this study calculated the numbers for those two regions that lead to the conclusion that there was not an issue in these two regions. Of course, routine monitoring of the debris in those arenas must be conducted with mitigation approaches in place for quick responses. They will be studied later along with how to leverage derelict GEO satellites as mass for apex anchors.

If there is a severance below 2,000 km, several operational procedures must be in place to ensure the safety of space elevator infrastructures. These ideas have been discussed but not studied. As such, they are listed here with recognition that analysis is a necessary action to be achieved within the near future:

1. Design an emergency response that unreels tether material from GEO downward when a large tension change warns that a severance has occurred.
2. Provide an emergency lowering of tether from 2,500 km upon severance from below in the highest probability areas (LEO high density orbits). This would be the case where there are no multiple legs, just a reel at 2,500 km to compensate for severance.

---

[14] Fitzgerald, M., "Space Elevator Architecture's Debris Mitigation Note #25," ISEC internal publications - available at www.isec.org, March 2019.

3. Provide multiple leg infrastructure from 2,500 km enabling transfer of the main stress to a replacement "principal tether."

Chapter 6: Conclusions for 2030

This position paper has shown the NASA space debris numbers of tracked items (>10 cms) for 2010 and 2019. In addition, it addresses the question of what will happen when new satellite constellations become operational with respect to the challenge for space elevators. The estimate shown for 2030 is a NASA number including the increase due to future satellite constellations. Even with this new number (essentially twice the current number) the conclusions are essentially the same as shown in the 2010 ISEC Study Report (see Appendix A, 2010 Study Conclusions). The following list uses the 2010 words and replaces the numbers for 2030.

- The geosynchronous (and above) region was not a significant threat.
- The MEO region has similarly low probability of conjunction.
- The LEO region is the area of major concern with the following insights:
  - Untracked, small (<10 cm) debris will impact a Space Elevator in (LEO 200-2000 km), on the average, once every **four** days; and therefore, must be designed for appropriate impact velocities and energies (was ten days with 2010 numbers).
  - Tracked debris will impact the total LEO segment (200 – 2000 km) once every **40** days or multiple times a year if **no movement actions are taken**. (was 100 days)
  - Tracked debris will only impact a single 60 km stretch of LEO space elevator, on the average, every **seven** years and every **three** years in the peak regions **if no movement actions are taken**. (was 18 and 5)

In addition, the summary from 2010 is still valid. It was:

Space debris mitigation is an engineering problem with definable quantities such as density of debris and lengths/widths of targets. With proper knowledge and good operational procedures, the threat of space debris is not a show stopper by any means. However, mitigation approaches must be accepted and implemented robustly to ensure that engineering problems are addressed.[15]

An earlier chart is shown that reflects the above words. The bottom line is there will be conjunction possibilities periodically, but the mitigation techniques are well studied and should enable operational successes in the later part of the 2030's. Space Elevator operations should have no issues operating in the world of robust space flight and interplanetary activities.

Item	*2010*	*2019*	*2030 Est.*	**Comment**
Total Tracked Debris by NASA (2010 & 2019 measured, 2030 estimated)	15378	19137	38,000	Assume Internet constellations will add many space objects by 2030
Threats in GEO region (possible conjunction)	0.0026 per year	0.005 per year	0.01 per year	Good operational procedures a must.
Threat in MEO region (possible conjunction)	0.0003 per year	0.0006 per year	0.0012 per year	Good operational procedures a must.
Untracked, small (<10 cm) debris will impact a Space Elevator in (LEO 200-2000 km), on the average;	Once every ten days	Once every 7.5 days	Once every 4 days	Design for tether and movement planned to account for this - with continuous repair[16]
Tracked debris will impact the total LEO segment (200 – 2000 km) if no actions are taken.	Once every 100 days or multiple times a year	Once every 75 days or several times a year	Once every 40 days or every two months or so	Note, this assumes there is no active movement of tracked objects or movement of the tether
Tracked debris will only impact a single 60 km stretch of LEO space elevator, on the average	Every 18 years with every 5 years in peak regions	Every 14 years with every 4 years in peak regions	Every 7 years with every 3 years in peak regions	Note, this assumes there is no active movement of tracked objects or movement of the tether

Table 2.6 Summary by Year and Altitude Region

---

[15] Swan, P., R. Penny, C. Swan, "Space Elevator Survivability Space Debris Mitigation," Publisher Lulu.com, 2010.
[16] Repair of tether from small debris impacts is a must in the design, development and operational phases

Chapter 7: Recommendations:

The basic recommendation for space elevators addressing the challenge of space debris is that the design and architectural teams should emphasize teamwork and collaboration. This is very consistent with the recommendations from the 2010 study (see Appendix B, 2010 Study Recommendations). Team work will help develop designs within both space elevator and space traffic management architectures that will work together to ensure safe transit of both orbiting space assets and the moving (but Earth attached) space elevator. Several of the concepts are shared below that can definitely improve future space operations.

The first set of concepts resides inside the arena of Architectural and Engineering Design Inputs:
- A multi-leg design to spread risk across multiple tethers.
- Designing tethers to survive small debris, ensuring that impact velocity and "shock" will not be catastrophic to tether integrity.
- Incorporate repair tether climbers that mend small holes or rips in the tether ensuring continuity of the tether over its lifetime.
- In addition to specific inputs to the design, the teams should support operational approaches developed for safe operations.

The second set of recommendations is centered around the necessity to collaborate with other operational space teams:

- Initially, the space elevator development team must establish co-operations with Space Traffic Management organizations to ensure that the transfer of information goes both ways and is almost instantaneous when there is a potential conjunction of assets. Warnings should be projected out to 24, 48 and 72 hours to allow space elevator and operational space systems to ensure they do not come close enough to threaten each other.

- Indeed, cleaning up our orbits is important so the coordination with owners of space assets (operational and derelicts), especially at

GEO, will benefit both organizations. These assets can be leveraged for many purposes once space elevators have enabled frequent and inexpensive access to those altitudes.

- Of course, the space elevator organization should assist with other organizations responsible for removal of space debris.

The third set of recommendations deals with Operational Approaches:

- One of the most productive and effective manners of mitigation of space debris is through passive approaches for debris mitigation. They include: a multi-leg design to spread the risk; varying tether design by altitude to respond to various characteristics such as gravity, wind and radiation; and, the growth option of multiple parallel tethers for greater carrying capacity.

- Considerations for an Active Approach are also mandatory. This would include: tether movement upon demand from a multitude of approaches (reel in-out/ change in climber movement/ movement of Earth Port or Apex Anchor); creation of a core element of the mitigation system - an on-orbit Sentry Satellite System for capture/movement/disposal of debris; and, of course, the evident need to develop an approach for recovery from severance of the tether.

Appendices:

Appendix A: 2010 Study Conclusions

Appendix B: 2010 Study Recommendations

Appendix C: ISEC Architecture Note 25,

Appendix D: ISEC Study Process and completed studies

Appendix E: International Space Elevator Consortium

Appendix A: 2010 Study Conclusions

Excerpt from year-long study entitled:
*"Space Elevator Survivability
Space Debris Mitigation"*

Chapter 5 – Conclusions

5.0   Debris Density Reduction

During the preparation for this pamphlet, it became apparent that the community of space debris experts is at a watershed year. They have convinced themselves that there is a need for more robust action than merely mandating mitigation techniques on rocket and spacecraft designs. At the December 2009 Space Debris Removal Conference sponsored by both NASA and DARPA, the majority agreed that space faring nations must do more than currently required (but unenforced), they must actually remove large debris from orbit. This was confirmed in Moscow (April 2010 conference) and Paris (June 2010 conference) with discussions on what types and sizes of debris must be removed, how many per year, and finally what impact would that have on the probability of collisions. The space elevator community endorses those efforts, but would like to encourage further actions to "improve the environment" by reducing density numbers.

5.1   Probability of Collisions.

Earlier in this pamphlet, the probability of collision for a 100,000 km space elevator with the debris density of April 2010 was calculated. Those numbers showed:

- The geosynchronous (and super GEO) region was not a significant threat of collision.
- The MEO region has similarly low probability of collision.
- The LEO region is the area of major concern with the following insights:
- Untracked, small (<10 cm) debris will impact a Space Elevator in (LEO 200-2000 km), on the average, once every ten days; and, therefore, must be designed for impact velocities and energies.
- Tracked debris will impact the total LEO segment (200 – 2000 km) once every 100 days or multiple times a year if no actions are taken.
- Tracked debris will only impact a single 60 km stretch of LEO space elevator, on the average, every 18 years and every five years in the peak regions.

5.2   Significant Questions:
In the first chapter, a few significant questions were asked to help identify the principle issues. They are represented here with the conclusions from the analyses.

Q. Does space debris cause concern for space elevator?
Answer: YES.
Q. How precisely does one need to know the location of the space elevator ribbon segments?
Answer: Estimate one meter (can be accomplished by GPS or ground based laser reflectors).
Q. How precisely does one have to know the location, and propagated location of large space debris?
Answer: Within 100 meters for 24 hours.
Q. What are the projected levels of concern and what needs to be accomplished prior to operations?
Answer: Knowledge of all tracked debris with improved propagation models and routine knowledge of ribbon location.
Q. How do we mitigate the risk of orbiting debris and satellites colliding with the space elevator?
Answer: Knowledge and planning.
Q. What is the probability of puncture from impacts of small items?
Answer: Close to 100%; therefore, it must be assumed in the design phase of the ribbon.
Q. What is the probability of severing by large orbiting objects?
Answer: Almost zero.

5.3   Conclusion

Space debris mitigation is an engineering problem with definable quantities such as density of debris and lengths/widths of targets. With proper knowledge and good operational procedures, the threat of space debris is not a show stopper by any means. However, mitigation approaches must be accepted and implemented robustly to ensure that engineering problems do not become a catastrophic failure event.

Appendix B: 2010 Study Recommendations

Excerpt from year-long study entitled:
"Space Elevator Survivability
Space Debris Mitigation"

Chapter 6 – Recommendations

6.0　Recommendations

Recommendations are divided into areas where they can be successfully implemented and will significantly improve the survivability of the Space Elevator vs. Space Debris. The conclusions lay out identifiable actions for the various communities.

6.1　Active Player Actions

6.1.1　Space Elevator Community
The space elevator community must lead the way in working with, and guiding, the space community. One of the first items would be to determine the best way to geolocate ribbon elements down to 100 meter segments to within one meter accuracy. The next item is to ensure the design of the ribbon is compatible with the environment. The current robust design is to have a one meter wide, woven ribbon that is tolerant to small debris penetrations [with, of course, a methodology for inspecting the ribbon and repairing in a timely manner]. Operational procedures must ensure that the ribbon element, whose location we know, will not be in the same location, at the same time, as a larger piece of debris (which can be tracked and its location projected).

6.1.2　Space Community
The space community must continue to improve its reporting and tracking of the environment,. They need to identify and implement programs to assure more precise tracking of debris in a timely manner. Another must is to improve the ephemeris propagation technologies so that the timeline for accuracy can be lengthened to a workable timeframe for the commercial world. The inclusion of GPS capabilities on all satellites as well as the ability of each to communicate to an operations center for timely updates of the database should be mandatory. One policy item that could significantly assist in the process would be the publication of the ephemerides of the debris/satellite in a timely [daily] manner. Another item would be a designation of a set of "rules of the road" so that all satellites could let all others know where they are and where they will be in the future (similar to commercial airliners). A current practice that should become mandatory is de-orbit of all LEO satellites within 25 years. And, of course, the biggest item is to immediately initiate a robust program to remove large debris from orbit [maybe ten items per year per participating country].

6.1.3　Satellite Launcher and Operator

Indeed, the operators of both launch vehicles and satellites must treat their environment in a manner that would encourage others to use the resource. If we are to have robust transportation to and from low Earth orbit, safety factors drive us to clean up space debris. In addition, as the number of assets in space increases, the probabilities of accidents, such as the IRIDIUM-Cosmos crash, increases. And finally, we encourage launch operators to consider how they can benefit from the use of space elevators to move payloads on a "real" transportation infrastructure. While they are considering the change, they can contemplate how a space elevator can make their tasks easier, cheaper, and safer, such as the removal of space debris.

6.2   Concluding Thoughts

The risk of collision of a tracked object with the space elevator is low; but, the consequences are high. Therefore, it must be addressed. Three quick thoughts should stimulate more discussions.

6.2.1   Multiple Space Elevators FIRST!

The primary mitigation technique is multiple ribbons. Once we overcome the gravity well we must ensure we always have a ribbon available to build another ribbon. The risk of collision with an untracked object is high but the consequences are low. Periodic "inspect and repair as necessary" by a repair robot should preserve the capability of individual ribbons. By immediately building the second, and then a third, the likelihood of losing operational space elevator access to orbit diminishes and humankind will never again be subject to the constraints of a gravity well.

6.2.2   Another Perspective – Steps Forward

When it comes to the international community, the general rule is that new owner/operators must not interfere with systems already in place (grandfathered). From a debris mitigation standpoint, it should be expected that space elevator owner/operators must not interfere with existing systems. Therefore, a space elevator should not pose a threat to current orbiting satellite systems. If we consider IRIDIUM (66 satellites in the 774-784 km band) and use the aforementioned formula, IRIDIUM would have a .055 PC with a single space elevator for a year. As a maneuver of a couple kilometers would almost certainly disable the use of their crosslinks, IRIDIUM would likely rather not perform such maneuvers. This requires the space elevator community to ensure their planned operations include collision avoidance activities that do not require existing systems to perform collision avoidance maneuvers.

6.2.3   Another Perspective – Enablers

What is currently not affordable in the space debris mitigation and removal will become easily achievable with inexpensive access to space through a space elevator infrastructure.

6.3   Aggressively Endorse Initiatives

As a space elevator concept comes of age, with a solid systems engineering program, three timely initiatives dealing with the space debris community are required:

Initiate Space Elevator Corridor
 "**Rules of the Road**" must be initiated to enable a space elevator vertical corridor to exist. Control of nodal passing must be implemented around the world with a mature set of rules ensuring that a space elevator can become a reality.
Initiate a De-Orbit Capability through A Prize Approach

Many papers and engineering concepts have surfaced that deal with elimination of current and future orbital debris. However, cost has always limited these activities to studies without follow-on engineering orbital tests. As a space elevator is funded and goes forward, investment in environmental cleanup should be included in all planning and funding requirements. One idea is to create a prize for the first organization to de-orbit a rocket body with a current estimated lifetime of ten years or more. The prize could be called the "Space Debris Enterprise Award." In addition, rewards can incentivize de-orbiting debris that is hazardous to the future of space elevators. New debris must become at least as socially, and perhaps legally, unacceptable as terrestrial pollution. Another approach is a space superfund as proposed in
http://www.popsci.com/technology/article/2010-12/new-report-calls-space-superfund-clean-junk-low-earth-orbit.

Go Beyond a "Zero Debris" Position
The International Academy of Astronautics has published a position paper on space debris.[17] In that paper the Academy takes the position that it is the goal of all space faring nations to create zero space debris within the three important regions. The LEO, navigation constellation ring, and GEO belt are identified. To ensure a healthy space elevator, the concept must be broadened to include all orbits. The mandatory implementation of Zero Debris Requirements would be early in a space systems design for programs prior to their Preliminary Design Reviews. However, the positive impact on a space elevator and other future initiatives will be tremendous. This pamphlet's concept would be to ensure that zero debris creation is implemented with a new goal of "improving the environment – not simply less pollution."

6.4   Final Recommendation
We hope that this study has raised the awareness of the problem to the space elevator stakeholders and all other users of the near Earth space environment. Further, we hope that this study will spur action to implement policies and directives to mitigate and reduce the risk of collision.

---

[17] Hussey, John ed., Paper on Space Debris Mitigation Guidelines for Spacecraft, Draft – International Academy of Astronautics, 2003.

International Space Elevator Consortium   **ISEC Position Paper # 2020-1**

Appendix C: Space Elevator Architecture Note #25, March 2019

"Space Elevator Architecture's Debris Mitigation Roles"

TOPICS:
- Debris alert ➔ Warning needs
- Debris sizing ➔ as a threat variant
- Space Elevator Tether Movement ➔ passive defense **Proposed**
- The Sentry System ➔ an Architecture adjunct
- System Recovery ➔ Post debris-event actions

Michael A. Fitzgerald
Senior Exec VP and Co-Founder
Galactic Harbour Associates, Inc
Space Elevator Transportation & Enterprise Systems

Personal Prolog

This is an Architecture Note. It is the opinion of the Chief Architect. It represents an effort to document ongoing science and engineering discussions. It is one of many to be published over time. Most importantly, it is a sincere effort to be the diary, or the chronicle, of the multitude of our technical considerations as we progress; along the pathway developing the Space Elevator.

Michael A. Fitzgerald

Space Elevator Summary Statement: Performance Attributes Debris Mitigation

It may be a myth, but old story goes that once upon a time (circa 1895) there were only two automobiles in the entire state of Ohio, and they ended up colliding with each other. So, urban myth or not; It probably is **NOT** a good idea to think that "collisions" will not affect your system. In our case, it is collisions with space debris and/or "rogue" satellites.

The Space Elevator Transportation System will soon be beginning its next development stage; engineering validation. In this stage, all needed capabilities are reviewed for engineering realism, development risk, performance projections, test data and operations simulations availability, and more. In most cases, the Space Elevator performance will be derived from subsystems to be designed, developed, and built within industry's Space Elevator Transportation System development program(s). In other cases, the performance capability sought will be provided by others; an entity not part of the development program. One such case will be "at large"

/ or "on the market" capabilities to resolve the real and potential threat from space debris and "rogue" space craft.

ISEC believes that debris mitigation concepts will be built, operating, and thriving before the Space Elevator Transportation System reaches operational status. To that end, this paper serves as the initial characterization of how the Elevator can allocate the needed performance to a system then available. That system would serve as Sentry; capturing, destroying, and / or removing the debris threat. Additionally, other "topics" must be addressed.

Debris Alert

ISEC foresees a close and interactive communication with the military Combined Space Operations Control Center; known familiarly as CSpOC. CSpOC is responsible for tracking the thousands of debris pieces and providing the orbital parameters of those pieces to operating space users. In addition, commercial capabilities have emerged which offer forming and formatting that information; operationally satisfying their commercial customers. Analytical Graphics, Inc.'s ExoAnalytic Solutions has been active in this regard for years.

At any rate, the Space Elevator team expects that the Sentry system operator will be able to depend on a warning forecast at 72 hours (tbr), of a convergence / close approach to a Space Elevator tether location (accuracy tbr). This closure accuracy will improve (improvement tbr) as convergence approaches. The Space Elevator team expects the commercial team will hold a place on the Debris Mitigation Chair in the Space Elevator's Headquarter's Primary Operations Center (HQ/POC). The Space Elevator team expects to share the elevators self-surveillance data and other location information with Space Situational Awareness authorities (tbd).

Debris Size

The Space Elevator team foresees that the Space Elevator tether will be able to withstand "collisions" of space debris when the debris is small (size tbd). That engineering character (e. g. size, mass, and speed) has not yet been assessed AND the operational tether maintenance concept is still being defined. It is expected that the Space Elevator Tether Climber will be able to detect tether "scars", and the Climbers are expected to have some level of minor tether repair.

In any case, the team expects the Debris Mitigation Chair to work with its Space Situational Awareness member to predict damage of an impending collision, assess the damage caused by collisions not predicted or that could not be avoided. Damage assessment is an imperative; pre & post event.

Space Elevator Tether Movement

The Space Elevator team has long cited the capability of the tether to move away from an impending collision. It is much like a simple "jump rope" movement; the movement generated by movement of the Earth Port's Tether Terminus with movement augmented by Reel In–Reel Out (RIRO) spools at the Earth Port and at the Apex. Simulation work is necessary but the impact of such a motion on tether / climber operations appears to be negligible. The team would rather have to "jump" rope only when necessary. In any event, the jump rope motion will be retained within our Debris Mitigation efforts.

The Sentry

The Space Elevator team has decided to examine an added capability within Debris Mitigation. In a concept called "Sentry", space debris will be intercepted and removed before a collision takes place. Debris headed for a collision with a portion of the Space Elevator will have intercept priority but, if otherwise not encumbered by that priority; the Sentry will gather and dispose of other space debris as a matter of course.

The team proposes that the Sentry be independently built as its engineering competencies are developed. Its initial operations are dictated by its schedule; not the Space Elevator's schedule. However, it should be operational as part of the Space Elevator Architecture as part of the Elevator's Tether deployment and build up. In our terms, by Sequence #4.

The team foresees a concept in which Sentry debris capture satellites would be in elliptical orbits along the lower tether; at least including the "debris belt" (tbr). The number of debris collectors needed will be determined based on a flight operations analysis; with periodicity and revisit established by the 72 hour (tbr) window and the number of captures needed to maintain mitigation.

Space Elevator Recovery Operations

The ISEC team has done little in this regard; but in the coming months ISEC will begin definition of Tether & Climber operations fashioned to minimize the impact of a tether break. A key aspect of recovery operations will be where the break occurs. In many cases, the lower portion and attached objects will reenter; being destroyed. In other cases, the altitude of the break would lead to the tether moving away from reentry and thus be accessible by the RIRO's. Additional RIRO's would be valuable in that circumstance. The economic value of the payloads in the several climbers affected by a break also makes recovery operations mandatory. (Dah!)

In closing, Space Elevator Outreach Program

Space debris is expected to be part of space operations for an extended period in this century. The real mitigation approach is the establishment of policy and actions that will prevent, or at least extensively reduce, the creation of debris in the first place.

This paper is ISEC's attempt to document the approach to mitigate the Space Elevator mission impact and safety issues when space debris meets the Space Elevator's tether.

We strongly suggest that other activities confront the issue. In the era of space debris, we all live in Ohio.   More to be supplied -- *Fitzer*

Appendix D: ISEC Completed Studies

List of International Space Elevator Study Reports Available on www.isec.org or purchase from www.lulu.com

Space Elevator Survivability, Space Debris Mitigation 2011

Space Elevator Concept of Operations 2013,

Design Considerations for Space Elevator Tether Climbers 2014,

Space Elevator Architectures and Roadmaps 2015,

Design Considerations for Space Elevator Earth Port 2016,

Design Considerations for Space Elevator Apex Anchor and GEO Node, 2017

Design Considerations for a Software Space Elevator Simulator, 2018

Design Considerations for a Multi-Stage Space Elevator, 2019

Today's Space Elevator, Status as of Fall 2019

Space Elevators: A History, 2017

Today's Space Elevator – A Status Report 2019

In the last year, the International Space Elevator Consortium assessed that the basic technological needs for the space elevator can be met with current capabilities: and, each segment of the Space Elevator Transportation System is ready for testing leading to engineering validation. Because of the availability of a new material as a potential Space Elevator tether, the community strongly believes that a Space Elevator will be initiated in the near term. Included in the book is a series of appendices that are tremendous references to the status of the space elevator today. Included are a lexicon of space elevator terms, over 750 references in the bibliography, short descriptions of eight ISEC year-long studies and two IAA 4-year studies on space elevators, as well as a summary of over 20 Architectural Notes covering the development of space elevator technologies.

This one document can bring the reader up to speed of the whole space elevator community across policy, technologies, developmental phases, management, and testing progress.

Get It Now! A study report explaining the Space Elevator Status - Fall 2019 with a Bibliography (over 750 inputs), a Lexicon (global agreement of terms), and Explanation of Studies (2 IAA & 8 ISEC). Download it now from the www.isec.org - or buy it on www.lulu.com.

Design Considerations for a Multi-Stage Space Elevator [2019]

Design Considerations for the Multi-stage Space Elevator

John M. Knapman
Peter Glaskowsky
Dan Gleeson
Vern Hall
Dennis Wright
Michael Fitzgerald
Peter A. Swan

To build a space elevator, the toughest challenge is to find material that is strong enough for a self-supporting tether. Building it in multiple stages is a way of overcoming that challenge. Using the concept of dynamically supported structures, it is possible to build upwards from the earth's surface and provide supports for the lowest parts of the tether, where gravity is strongest. A five-stage design would support a tether made of carbon fiber yarn that is commercially available today. A two-stage design can support a tether with less than one-third of the strength previously thought necessary. The study report analyses the proposal in detail, covering the underlying physics and technology, design options and prototyping work. Authors: John M. Knapman, Peter Glaskowsky, Dan Gleeson, Vern Hall, Dennis Wright, Michael Fitzgerald, Peter Swan

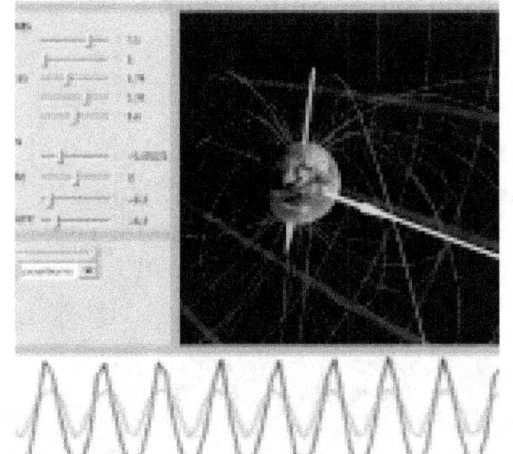

Design Considerations for a Software Space Elevator Simulator [2018]

This study report gives a detailed analysis of all the design considerations for a Software Space Elevator Simulator. From the Executive Summary: As with all large, modern engineering projects, detailed computer simulations of the space elevator will be essential during its design, construction and operational phases. Within the context of these phases, this study enumerated 14 use cases which the simulation software must address, ranging from 3D dynamics and electrodynamics calculations of space elevator motion, to the effects of payload capture and release at various points along the tether, to the effects of friction arising from the interaction of the space elevator climber with the tether. Proceeding from these use cases, requirements were imposed on the software design and an outline for its development was sketched. Authors: Dennis H. Wright, Steven Avery, John Knapman, Martin Lades, Paul Roubekas, Pete A. Swan

Design Considerations of a Space Elevator Apex Anchor and GEO Node [2017]

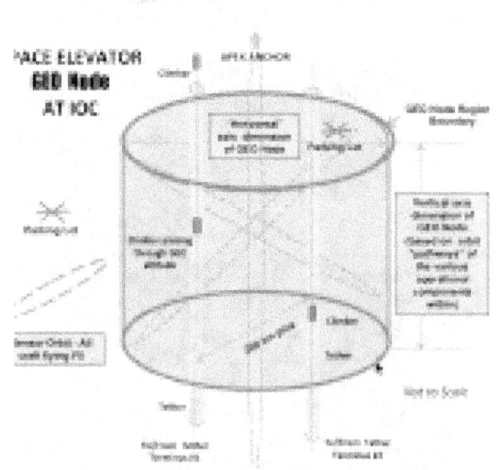

This year, ISEC chose to address the design considerations for the Apex Anchor and Geosynchronous Earth Orbit (GEO) Node. As was discussed in the Architectures and Roadmap report, ISEC understands where the technologies are today and where we would like them to be in order to reach Initial Operational Capability (IOC). The goal of this study team is to add to the "body of knowledge" relative to the two topics addressed herein. To ensure complete understanding during this study report, the following definitions were developed: Space Elevator Column, Earth Port & Earth Port Region, GEO Node & GEO Region, and Apex Anchor & Apex Anchor Region. In addition, the needs [functional requirements] were discussed for each of those regions and complexes. Throughout the text, the initial destination is described as the IOC for the Space Transportation System. The Space Elevator Transportation System is comprised of one Earth Port with two tether termini, multiple Apex Anchors supporting 100,000 km Tethers, 14 Tether Climbers, and a single Headquarters and Primary Operations Center. The GEO Node supports the Space Elevator Transportation System with a range of "overhead' functions; e. g. test, safety, and support. Authors: Michael Fitzgerald, Vern Hall, Peter Swan, and Cathy Swan.

Design Considerations of a Space Elevator Earth Port [2016]

This study report provides the International Space Elevator Consortium's (ISEC) view of the Earth Port (formerly known as the Marine Node) of a Space Elevator system. The Earth Port: Serves as a mechanical and dynamical termination of the space elevator tether; Serves as a port for receiving and sending Ocean Going Vessels (OGVs); Provides landing pads for helicopters from the OGVs; Serves as a facility for attaching and detaching payloads to and from tether climbers and attaching and detaching climbers to and from the tether; Provides tether climber power for the 40 km above the Floating Operations Platform (FOP); and, Provides food and accommodation for crew members as well as power, desalinization, waste management and other such support.
Authors: Robert E. 'Skip' Penny, Jr, Vern Hall, Peter Glaskowsky, and Sandee Schaeffer.

Space Elevator Survivability, Space Debris Mitigation [2011]

This report focuses on the issue of Space Debris in relation to a Space Elevator. Many people looking at the idea of a Space Elevator for the first time are concerned about how the ever-growing problem of Space Debris will affect it. This report gives an honest look at the numbers, where the Space Elevator is most vulnerable and what can be done about the problem. It shows that space debris is a manageable problem, giving proper foresight and engineering. Authors: Dr. Peter Swan, Cathy Swan and Robert "Skip" Penny.

Space Elevator Concept of Operations [2013]

This report describes and discusses a plausible Operations scenario for a Space Elevator. This report addresses initial commercial operations of a space elevator pair with robotic climbers. This report has been developed to help define a starting point for an initial space elevator infrastructure. It is assumed that there are two space elevators in place to ensure continuation of our escape from the gravity well. It also assumes that a sufficient number of climbers are available for delivering of spacecraft and other payloads to orbit, and, if required, return them to earth. In addition, this report is designed to be the initial operations concept from which many improvements will occur as future knowledge and experience drives infrastructure concept revisions.
Authors:
Dr. Peter Swan, Cathy Swan and Robert "Skip" Penny.

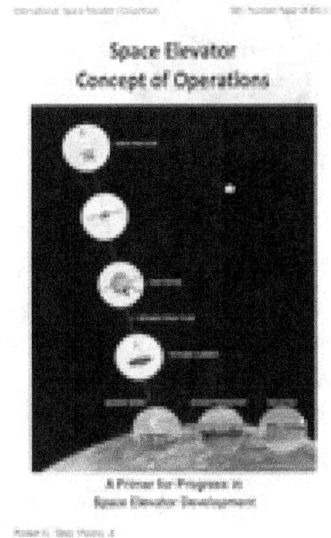

Design Considerations for Space Elevator Tether Climbers[2014]

The subject selected for this 2013 study is the Space Elevator Tether Climber. The objective of the one year study was to survey current concepts and technologies related to tether climbers, identify critical issues, questions, and concerns, assess their impact on the development of space elevators, and project towards the future. Authors: Dr. Peter Swan, Cathy Swan, Robert "Skip" Penny, John Knapman and Peter Glaskowsky.

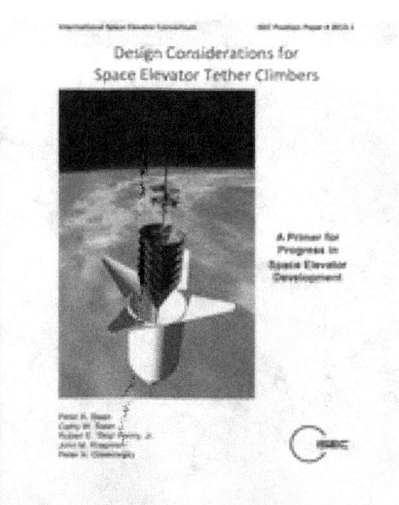

Space Elevator Architecture and Roadmaps –

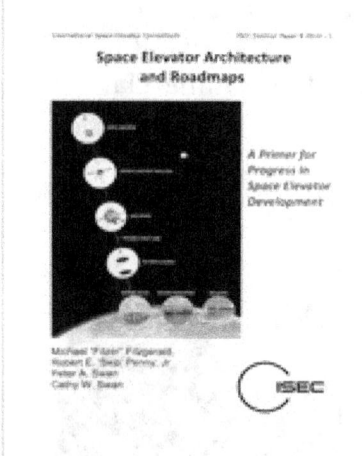

This 2014 study report establishes a baseline roadmap for designing space elevators for the future. This study addresses critical aspects of space elevator infrastructures: basic architectures and how we will get there with a roadmap. The roadmaps will leverage desired paths to lower risks and identify approaches for pulling together the diverse concepts. The three architectures in the literature today are solid looks at various approaches, while not providing that key element of "how will we get there?" Each path from today to the successful implementation of a space elevator infrastructure must be identified and discussed with respect to hurdles and milestones.

Authors: Michael "Fitzer" Fitzgerald, Peter Swan, Cathy Swan and Robert "Skip" Penny. Publication date: April, 2015

Appendix E: Description of International Space Elevator Consortium (www.isec.org)

Who We Are
The International Space Elevator Consortium (ISEC) is composed of individuals and organizations from around the world who share a vision of humanity in space.

Our Vision
A world with inexpensive, safe, routine, and efficient access to space for the benefit of all mankind.

Our Mission
The ISEC promotes the development, construction and operation of a space elevator infrastructure as a revolutionary and efficient way to space for all humanity.

What We Do
- Provide technical leadership promoting development, construction, and operation of space elevator infrastructures.
- Become the "go to" organization for all things space elevator.
- Energize and stimulate the public and the space community to support a space elevator for low cost access to space.
- Stimulate science, technology, engineering, and mathematics (STEM) educational activities while supporting educational gatherings, meetings, workshops, classes, and other similar events to carry out this mission.

A Brief History of ISEC
The idea for an organization like ISEC had been discussed for years, but it wasn't until the Space Elevator Conference in Redmond, Washington, in July of 2008, that things became serious. Interest and enthusiasm for a space elevator had reached an all-time peak and, with Space Elevator conferences upcoming in both Europe and Japan, it was felt that this was the time to formalize an international organization. An initial set of

directors and officers were elected and they immediately began the difficult task of unifying the disparate efforts of space elevator supporters worldwide.

ISEC's first Strategic Plan was adopted in January of 2010 and it is now the driving force behind ISEC's efforts. This Strategic Plan calls for adopting a yearly theme to focus ISEC activities. Because of our common goals and hopes for the future of mankind off-planet, ISEC became an Affiliate of the National Space Society in August of 2013. In addition, ISEC works closely with the Japanese Space Elevator Association.

Our Approach

ISEC's activities are pushing the concept of space elevators forward. These cross all disciplines and encourage people from around the world to participate. The following activities are being accomplished in parallel:

- Yearly conference – International space elevator conferences were initiated by Dr. Brad Edwards in the Seattle area in 2002. Follow-on conferences were in Santa Fe (2003), Washington DC (2004), Albuquerque (2005/6 –smaller sessions), and Seattle (2008 to the present). Each of these conferences had multiple discussions across the whole arena of space elevators with remarkable concepts and presentations.
- Yearlong technical studies – ISEC sponsors research into a focused topic each year to ensure progress in a discipline within the space elevator project. The first such study was conducted in 2010 to evaluate the threat of space debris. The products from these studies are reports that are published to document progress in the development of space elevators. They can be downloaded at www.isec.org.
- International Cooperation – ISEC supports many activities around the globe to ensure that space elevators keep progressing towards a developmental program. International activities include coordinating with the two other major societies focusing on space elevators: the Japanese Space Elevator Association and EuroSpaceward. In addition, ISEC supports symposia and presentations at the International Academy of Astronautics and the International Astronautical Federation Congress each year.
- Publications – ISEC publishes a monthly e-Newsletter, its yearly study reports and an annual technical journal [CLIMB] to help spread

information about space elevators. In addition, there is a magazine filled with space elevator literature called Via Ad Astra.
- Reference material – ISEC is building a Space Elevator Library, including a reference database of Space Elevator related papers and publications. (see section before this on references)
- Outreach – People need to be made aware of the idea of a space elevator. Our outreach activity is responsible for providing the blueprint to reach societal, governmental, educational, and media institutions and expose them to the benefits of space elevators. ISEC members are readily available to speak at conferences and other public events in support of the space elevator. In addition to our monthly e--Newsletter, we are also on Facebook, Linked In, and Twitter.
- Legal – The space elevator is going to break new legal ground. Existing space treaties may need to be amended. New treaties may be needed. International cooperation must be sought. Insurability will be a requirement. Legal activities encompass the legal environment of a space elevator - international maritime, air, and space law. Also, there will be interest within intellectual property, liability, and commerce law. Starting work on the legal foundation well in advance will result in a more rational product.
- History Committee – ISEC supports a small group of volunteers to document the history of space elevators. The committee's purpose is to provide insight into the progress being achieved currently and over the last century.
- Research Committee – ISEC is gathering the insight of researchers from around the world with respect to the future of space elevators. As scientific papers, reports and books are published, the research committee is pulling together this relative progress to assist academia and industry to progress towards an operational space elevator infrastructure.
- Competitions – ISEC has a history of actively supporting competitions that push technologies in the area of space elevators. The initial activities were centered on NASA's Centennial Challenges called "Elevator: 2010." Inside this were two specific challenges: Tether Challenge and Beam Power Challenge. The highlight came when Laser Motive won $900,000 in 2009, as they reached one kilometer in altitude racing other teams up a tether suspended from a helicopter. There were also multiple competitions where different strengths of

materials were tested going for a NASA prize – with no winners. In addition, ISEC supports the educational efforts of various organizations, such as the LEGO space elevator climb competition at our Seattle conference. Competitions have also been conducted in both Japan, Israel, and Europe.

ISEC is a traditional not-for-profit 501 (c) (3) organization with a board of directors and four officers: President, Vice President, Treasurer, and Secretary. inbox@isec.org / www.isec.org